MW00913342

We Want a Pet

by Winston White
illustrated by Molly Windsor

Harcourt
SCHOOL PUBLISHERS

Printed in China

ISBN-13: 978-0-15-358409-1
ISBN-10: 0-15-358409-2

Ordering Options
ISBN 10: 0-15-358356-8 (Grade K On-Level Collection)
ISBN 13: 978-0-15-358356-8 (Grade K On-Level Collection)
ISBN 10: 0-15-360661-4 (package of 5)
ISBN 13: 978-0-15-360661-8 (package of 5)

4 5 6 7 8 9 10   0940   15 14 13 12 11 10 09

We want to get a pet.

Look at this pet.
This pet can get wet.

Look at this pet.
This pet can hop.

Look at this pet.
It is not big.

Look at this pet.
This pet has a bell.

We want this pet.

This pet can wag!